Graphic Organizers in Science™

Learning About Earth's Cycles with Graphic Organizers

Isaac Nadeau

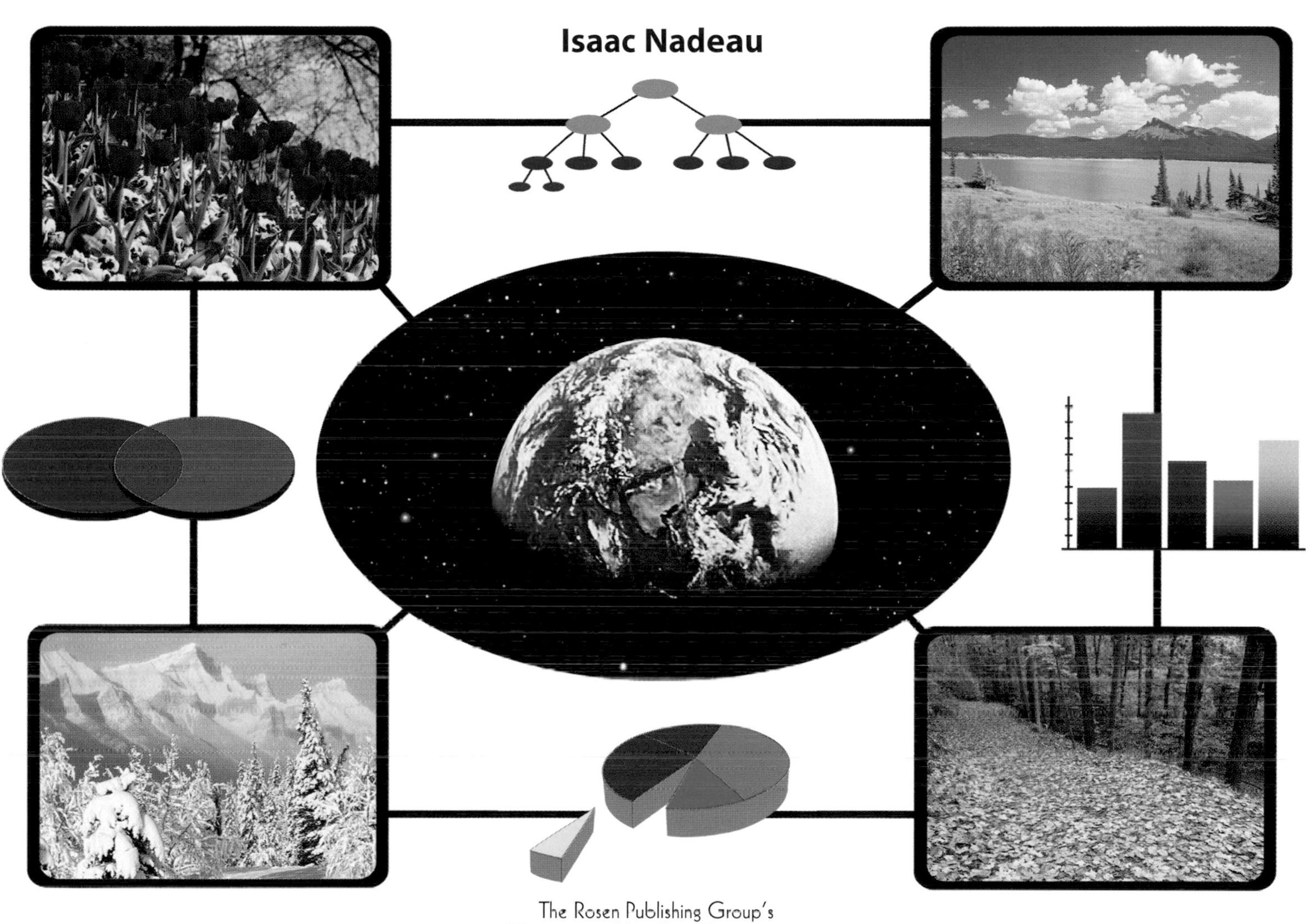

The Rosen Publishing Group's
PowerKids Press™
New York

For Jenny, and the Sun

Published in 2005 by The Rosen Publishing Group, Inc.
29 East 21st Street, New York, NY 10010

First Edition

Editor: Natashya Wilson
Book Design: Mike Donnellan

Photo Credits: Cover (center), pp. 15, 16 © DigitalVision; cover (top left) © CORBIS; cover (top right), (bottom left), (bottom right) © EyeWire; p. 4 © Susan Steinkammp/CORBIS; p. 7 © NASA/CORBIS; p. 11 © AFP/CORBIS; p. 12 (left) © Richard Hamilton Smith/CORBIS; p. 12 (right) © Richard Cummins/CORBIS.

Library of Congress Cataloging-in-Publication Data

Nadeau, Isaac.
Learning about earth's cycles with graphic organizers / Isaac Nadeau.— 1st ed.
v. cm. — (Graphic organizers in science)
Contents: What is a cycle? — Day and night — Once around the sun — Weather cycles — Freeze and thaw — Our traveling moon — Eclipses — Tides — Telling time — Earth's neighbors.
ISBN 1-4042-2807-1 (lib. bdg.)
ISBN 1-4042-5044-1 (paperback)
6-pack ISBN 1-4042-5045-X

1. Time—Juvenile literature. 2. Cycles—Juvenile literature. [1. Cycles. 2. Time. 3. Nature study.] I. Title. II. Series.
QB209.5.N33 2005
529—dc22

2003017269

Manufactured in the United States of America

Contents

Cycle: A Student's Daily Cycle

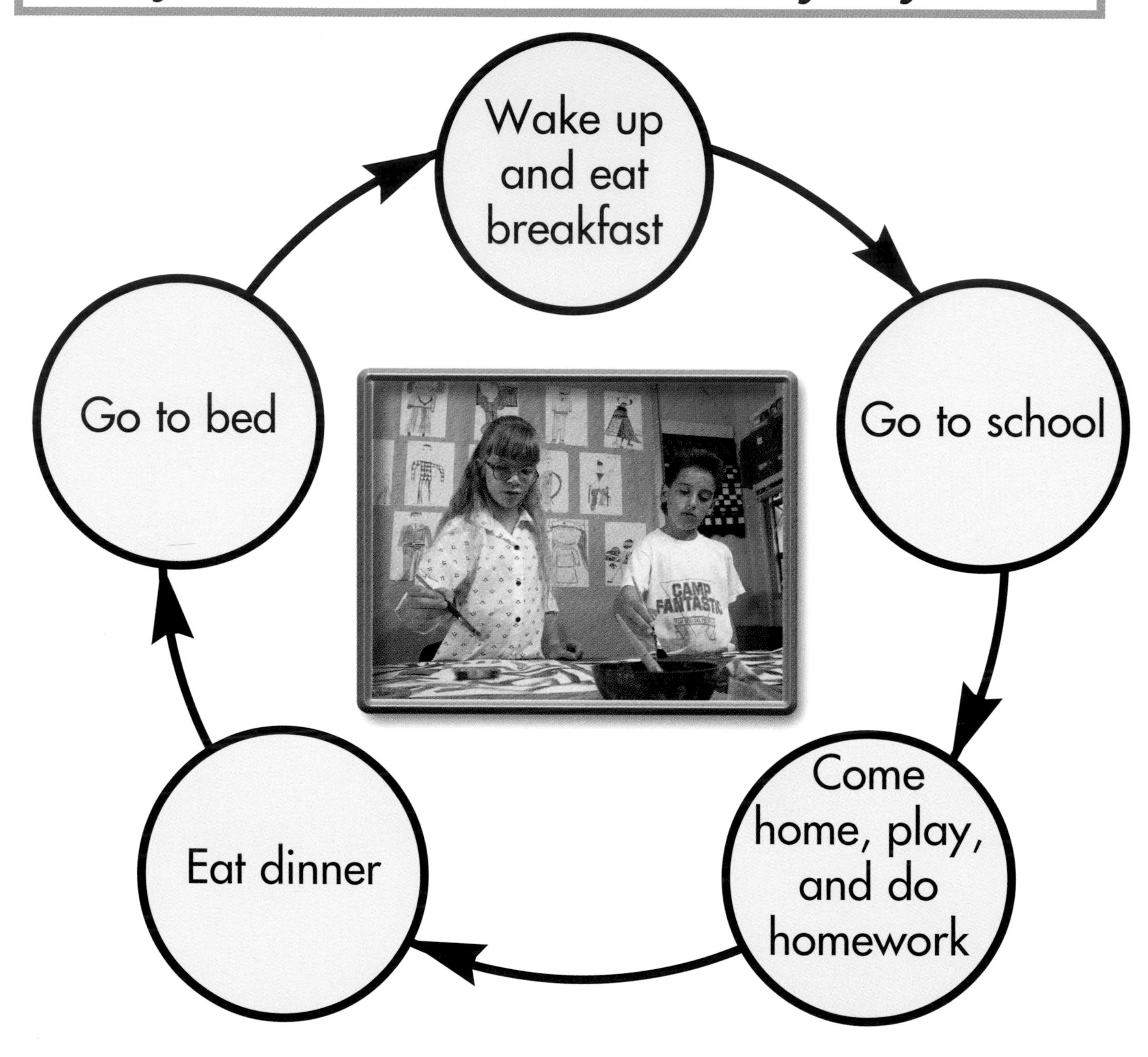

What Is a Cycle?

A cycle is a pattern that happens over and over again. You are part of many cycles every day. Waking up in the morning is part of a cycle. When you get out of bed, eat breakfast, and get ready for the day, you are at the beginning of a daily cycle. When you lie down again to sleep at night, the cycle continues. It starts again in the morning. Earth is part of many cycles, too. In some cases these cycles help us to **predict** what will happen in the **future**. The change of seasons is a cycle. Even in the coldest winter, we know that spring will come again. Learning about Earth's cycles can teach people a lot about the world.

Graphic organizers are tools that can help to make hard ideas clear. They arrange ideas in ways that make sense. In this book, graphic organizers are used to help explain Earth's cycles.

School is part of a student's weekday cycle. This graphic organizer, called a cycle, shows a simple outline of what one student's daily cycle is like. A cycle shows events that happen in the same order again and again. Putting continuing events into a cycle can help you to memorize and understand the order in which the events happen.

Day and Night

Earth is a giant **sphere**, or ball, that **rotates** around its **axis**. The axis is an imaginary line that runs through Earth from the North Pole to the South Pole. Earth takes 24 hours to turn in one full circle. This 24-hour daily cycle includes a day and a night. When you watch the sunrise in the morning, the place where you live is turning toward the Sun. As Earth turns, the Sun seems to move from east to west, getting higher in the sky. Around noon the Sun reaches its highest point in the sky. At sunset the place you live is turning away from the Sun. The sky becomes dark. When it is daytime on one side of Earth it is nighttime on the opposite side, because the Sun's light only reaches one half of Earth. The opposite half of Earth is in shadow.

Top: *This picture of Earth was taken near the Moon.* Bottom: *This graphic organizer is called a KWL chart. Making a KWL chart can help you to find out what you already know, what you want to know, and what you learn from studying a subject. This KWL chart shows some things you can learn by studying the daily cycle.*

KWL Chart: Day and Night

What I Know	What I Want to Know	What I Have Learned
• Day is light and night is dark.	• Why does day turn into night?	• The Sun shines on half of Earth. As Earth spins, places move from the light half into the dark half.
• Darkness makes it hard to see at night.	• Why is night dark?	• Nighttime happens in the parts of Earth that are turned away from the Sun. Earth's shadow creates nighttime.
• Days are usually warmer than nights.	• Why is it usually hotter at noon than at 7:00 A.M.?	• Around noon, a place is closer to the Sun than it was in the morning, so the Sun's rays hit it more directly and make it warmer.
• There are 24 hours in one day and night cycle.	• Why does the Sun rise in the east and set in the west?	• Earth spins eastward, so places to the east see the Sun before places west of them. Because Earth spins eastward, the Sun seems to move across the sky from east to west.

Concept Web: Seasons

SUMMER
The longest day of the year occurs around June 21. Summer begins. The North Pole tilts toward the Sun. Days are longer than nights. Temperatures are hot. Plants and grass wilt. Some animals come out only at night.

SPRING
Around March 21, day and night are of equal length. Spring begins. The North Pole begins to tilt toward the Sun. Days become longer than nights. Temperatures get warmer. Flowers bloom and many baby animals are born.

SEASONS
Earth spins at an angle to the Sun. This tilt causes the seasons as Earth travels in its orbit around the Sun.

FALL
Around September 22, day and night are of equal length. Fall begins. The North Pole begins to tilt away from the Sun. Days become shorter than nights. Temperatures get cooler. Leaves change color and fall off trees.

WINTER
The shortest day of the year occurs around December 22. Winter begins. The North Pole tilts away from the Sun. Days are shorter than nights. Temperatures are cold. Snow falls in northern and mountainous places.

Once Around the Sun

As Earth turns on its axis, it travels in a circle around the Sun. This circle is called an **orbit**. Earth takes about 365 days, or one year, to travel once around the Sun. As Earth orbits the Sun, the seasons change in a yearly cycle. Seasons are caused by the tilt, or lean, of Earth's axis. Earth can be divided into northern and southern halves called **hemispheres**. Each hemisphere is tilted toward the Sun for half of the year. When it is winter in the Northern Hemisphere, it is summer in the Southern Hemisphere. The hemisphere that is tilted toward the Sun gets more sunlight. The warmth makes the seasons change from winter to spring and then from spring to summer. As Earth continues around the Sun, the hemisphere gets less sunlight. Days grow shorter and colder. Seasons change from summer to fall and from fall to winter.

This graphic organizer is called a concept web. Concept webs are used to organize facts about a topic. The topic goes in the middle, and the facts go around it. Here the topic is seasons. The facts explain the features of the four seasons as they occur in the Northern Hemisphere, including in the United States. The pictures show a scene from each season.

Weather Cycles

Earth's movements cause daily and seasonal weather cycles. Wherever you live, you can observe patterns in the weather. These patterns make up the **climate** of your area. Changes in **temperature** are the easiest part of weather cycles to study. During the daytime, sunlight hits Earth. Temperatures rise. As night falls, temperatures drop. This cycle happens every day. Temperatures also change with seasons. During summer, the air is usually warm or hot. **Precipitation** falls as rain. The air is colder in winter, when days are short and there is not as much sun. In many places precipitation falls as snow. Wind is also part of seasonal weather cycles. For example, **monsoons** are winds that change direction in the summer and the winter. They bring heavy rain in the summer, when they blow inland from the ocean.

This graphic organizer is called a sequence chart. Sequence charts show how one step of an event leads to another. This chart shows the steps of monsoon season in the country of India. Warm summer temperatures cause ocean water to rise into the air as water vapor. The vapor forms rain clouds. Monsoons blow the clouds over India, where it rains.

Sequence Chart: Monsoon Season in India

India moves into spring and summer and temperatures rise.

↓

The monsoon winds change direction and blow in from the ocean instead of down from the northern lands.

↓

The monsoons blow the rain clouds that form above the Indian Ocean over India.

↓

Heavy rains fall.

In August 2002, the Indian city of Cochin was flooded after monsoons brought heavy rains that lasted for 48 hours. Here, people wade through the floodwaters.

Compare/Contrast Chart: Freeze-Thaw Cycle	Freeze	Thaw
Temperature	Temperatures are below 32°F (0°C).	Temperatures are above 32°F (0°C).
Water	Water becomes solid ice and takes up more space, making cracks in rocks larger.	Ice melts and turns back into water. The water flows into streams, lakes, and oceans as well as down into the ground.
Effect on Plant Growth	Plant growth slows and stops as the plant roots stop taking in water.	Plant roots take in water and plants begin to grow again.

Freeze and Thaw

The freeze-**thaw** cycle occurs as water on Earth is cooled and heated by changing temperatures. Water is one of the only **substances** that commonly can be found on Earth's surface as a liquid, a solid, and a gas. At 32°F (0°C), water freezes into ice. Water **expands**, or takes up more space, when it freezes. If water is in a crack in a rock when it freezes, it can expand enough to make a larger crack or to break the rock apart. This process also causes cracks in sidewalks and roads.

When water in the ground is frozen, the roots of plants cannot draw up water. Plant growth slows down or stops. When temperatures rise above 32°F, usually in spring, the ice thaws and melts. Once the ice melts, plant roots take in water again and the plants grow.

Top Left: *Ice makes cracks in rocks.* Top Right: *Once the ice melts, plants can grow in the cracks.* Bottom: *This graphic organizer is called a compare/contrast chart. A compare/contrast chart allows you to compare the features of different things. The topics are at the top of the columns. The features being compared are in the left column.*

Our Traveling Moon

The Moon orbits Earth just as Earth orbits the Sun. The light of the Sun casts a shadow on half of the Moon. As the Moon orbits Earth, our view of this shadow changes, and the Moon seems to change shape. The Moon takes about 29 ½ days to travel once around Earth. This cycle is called the **lunar** cycle. A lunar cycle has four main **phases**. During the new-moon phase, the Moon is between Earth and the Sun. The side of the Moon that faces Earth is dark. About one week later, at the first-quarter moon, the right half of the Moon is lit up. A full moon occurs when Earth is between the Moon and the Sun. Sunlight streams around Earth and lights up the whole face of the Moon. During the last-quarter moon the Moon's left half is lit up. One week later the lunar cycle begins again with a new moon.

Top: *This diagram shows the Moon orbiting Earth. Sunlight always hits the Moon from the same way, but our view of the Moon changes.* Bottom: *This is a chart about the lunar cycle. The left column lists the phases of the cycle. The information to the right of each phase gives more facts about the phase. The type of fact appears at the top of each column.*

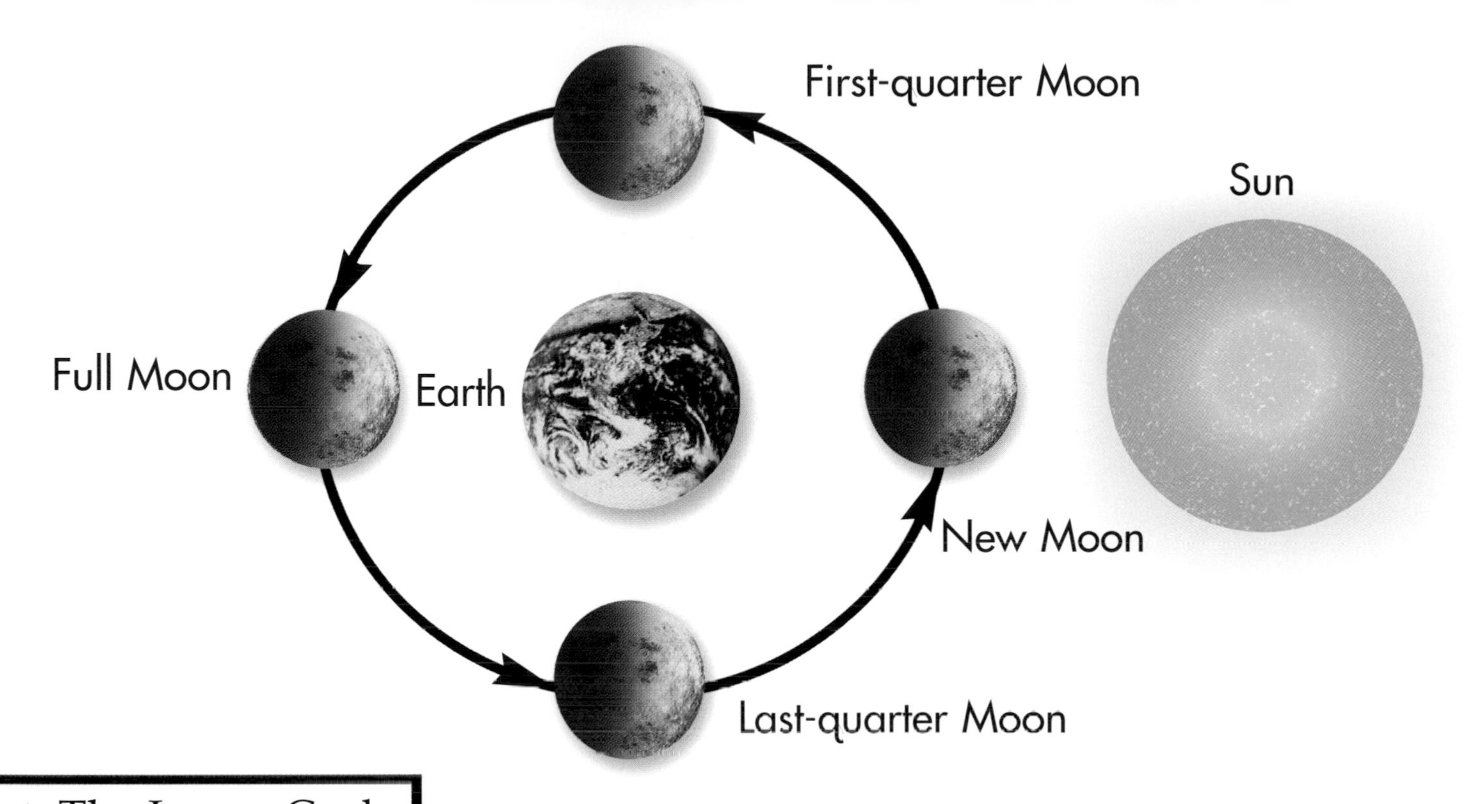

Chart: The Lunar Cycle

Phase of the Moon	Position of the Moon	What the Moon Looks Like from Earth	Moon
New moon	The Moon is between Earth and the Sun.	Dark. The light half of the Moon faces away from Earth.	
First-quarter moon	The Moon is at a right angle to Earth and the Sun.	Half full. The right side of the moon is lit up.	
Full moon	Earth is between the Moon and the Sun.	Full. The light half of the Moon faces Earth.	
Last-quarter moon	The Moon is at a right angle to Earth and the Sun.	Half full. The left side of the moon is lit up.	

Venn Diagram: Eclipses

Solar Eclipse	Both	Lunar Eclipse
• The Moon is between Earth and the Sun.	• Caused by the Moon's orbit around Earth.	• The Moon is on the other side of Earth from the Sun.
• The view of the Sun from Earth is blocked.	• The Sun, Earth, and the Moon line up.	• The Moon turns dark red.
• The Sun looks black.		• The Moon is in Earth's shadow.
• Happens only during a new moon.		• Happens only during a full moon.

Eclipses

Eclipses are part of the lunar cycle and the yearly cycle. They occur when Earth, the Moon, and the Sun line up. There are two kinds of eclipses. A lunar eclipse happens during a full moon when the Moon moves through Earth's shadow at night. For about 2 hours the Moon darkens. This happens up to three times per year. A solar eclipse occcurs during a new moon when the Moon moves directly between the Sun and Earth during the day. The Moon blocks the Sun, and the Moon's shadow falls on Earth. The day becomes dark for almost 8 minutes. Solar eclipses happen from two to five times per year. They can be seen only from the places where the Moon's shadow touches Earth. Eclipses do not happen during every full and new moon. The Moon only lines up with Earth and the Sun during those phases a few times per year.

Top: *This picture shows a solar eclipse.* Bottom: *This graphic organizer is called a Venn diagram. Venn diagrams show how two things are alike and how they are different. The features in the outer part of each circle apply to only one subject. The features in the middle, where the circles overlap, are shared by both subjects.*

Tides

The daily and lunar cycles cause Earth's oceans to rise and fall in cycles called tides. Tides are the result of the pull of **gravity**. The gravities of both the Sun and the Moon affect Earth's tides. However, the Moon's gravity has a greater effect because the Moon is much closer to Earth. The Moon's gravity pulls the water in the oceans toward it, so the water level rises on the part of Earth that is near the Moon. The Moon's gravity also pulls Earth away from the water on the opposite side of Earth. The water level rises there as well. As Earth rotates in its daily cycle, water levels rise in places as they pass through these two areas. This causes most places to have two high tides per day. Low tides occur between the high tides as places pass between the two areas. Most places have two low tides per day.

A table (top right) *organizes numeric facts. This table lists the times and the heights of the low and high tides near New York City on July 1, 2003. A line graph* (bottom) *shows how information changes over time. This line graph shows the changing tide level near New York City on July 1, 2003.*

Table: High and Low Tides, New York, NY	
Time	Height (feet)
4:33 A.M.	0.1
10:29 A.M.	4.2
4:22 P.M.	0.6
10:09 P.M.	5.3

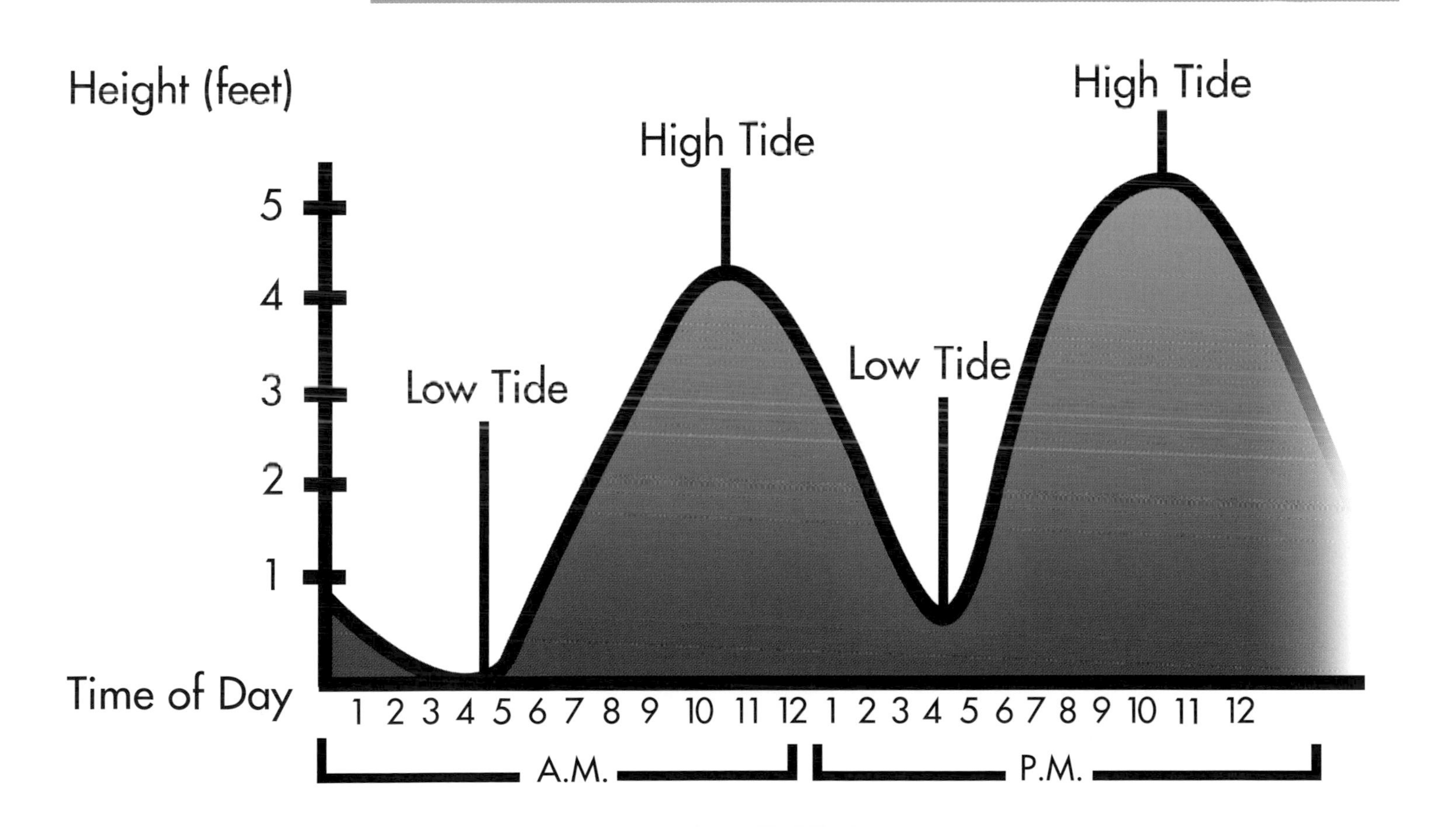

Timeline: A History of Telling Time

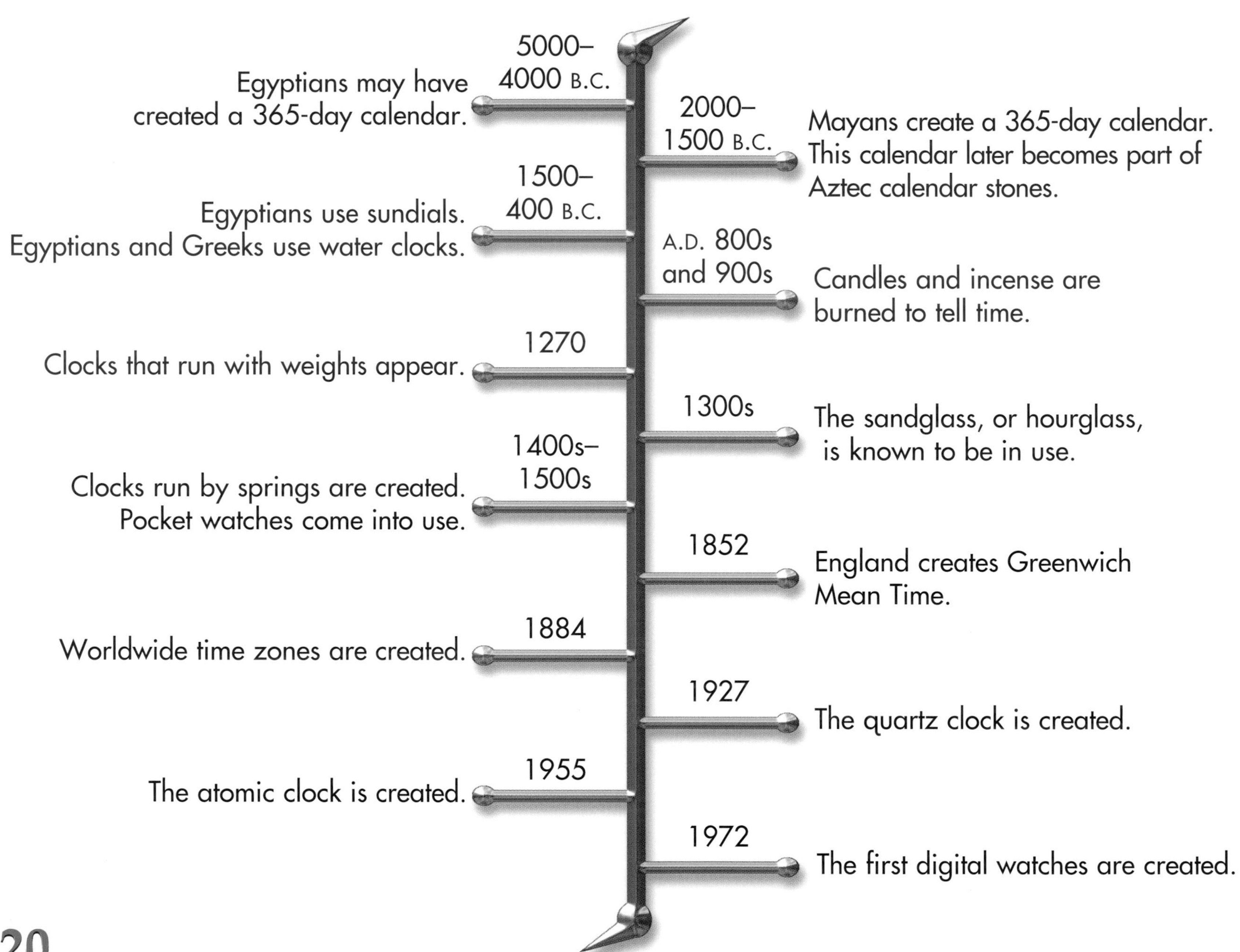

Telling Time

Throughout history Earth's cycles have helped people to mark time. The day comes from the daily cycle. The month comes from the lunar cycle. The year comes from Earth's journey around the Sun. People may have created 365-day calendars by 4000 B.C. Later, people used the movement of shadows to tell time during the day. The first **mechanical** clocks appeared in the 1300s. In 1884, people separated Earth's surface into 24 time zones, one for each hour. The time zones start at the prime meridian, an imaginary line from the North Pole to the South Pole, which passes through Greenwich, England. On the opposite side of Earth from the prime meridian is the international date line, which separates one day from the next. When it is noon at the prime meridian, it is midnight at the international date line.

This graphic organizer is called a timeline. A timeline helps you to remember when events happened as well as the order in which they happened. On this timeline, the earliest event is at the top. The most recent event is at the bottom.

Earth's Neighbors

Earth and the Moon are part of cycles in a bigger system called the solar system. The word "solar" comes from the Latin word *sol*, meaning "sun." The solar system includes everything that is in orbit around the Sun. There are nine planets, including Earth, in the solar system. All of the planets move in cycles. The planets' orbits around the Sun can be tracked from Earth with telescopes. Most of these planets have moons, which also move in cycles.

Our solar system is a tiny part of a **galaxy** called the Milky Way. It takes the Sun 225 million years to complete one full circle around the Milky Way. There are **billions** of stars like the Sun in the Milky Way. The Milky Way is one of billions of galaxies in space. As you wake up and go to school each day and see the seasons change each year, you are part of the many cycles on Earth and the billions of cycles in the **universe**.

Glossary

axis (AK-sis) A straight line around which an object turns or seems to turn.
billions (BIL-yunz) Thousands of millions. One billion is 1,000 millions.
climate (KLY-mit) The kind of weather a certain area has.
eclipses (ee-KLIPS-ez) Darkenings of the Sun or the Moon that occur when the light of the Sun is blocked by the Moon or when the light of the Moon is blocked by Earth's shadow.
expands (ek-SPANDZ) Spreads out or grows larger.
future (FYOO-chur) The time that is coming.
galaxy (GA-lik-see) A large group of stars and the planets that circle them.
graphic organizers (GRA-fik OR-guh-ny-zerz) Charts, graphs, and pictures that sort facts and ideas and make them clear.
gravity (GRA-vih-tee) The natural force that causes objects to move toward one another.
hemispheres (HEH-muh-sfeerz) Halves of Earth or another sphere.
lunar (LOO-ner) Of or about the Moon.
mechanical (meh-KA-nih-kul) Run by a machine or a tool.
monsoons (mon-SOONZ) Strong winds that change direction seasonally.
orbit (OR-bit) A circular path.
phases (FAYZ-ez) The different stages of the Moon as seen from Earth.
precipitation (preh-sih-pih-TAY-shun) Any moisture that falls from the sky.
predict (prih-DIKT) To make a guess based on facts or knowledge.
rotates (ROH-tayts) Turns in a circle.
sphere (SFEER) An object that is shaped like a ball.
substances (SUB-stan-siz) Any matter that takes up space.
temperature (TEM-pruh-cher) How hot or cold something is.
thaw (THAH) To turn from solid into liquid, such as when ice melts.
universe (YOO-nih-vers) Everything in space.

Index

Web Sites

Due to the changing nature of Internet links, PowerKids Press has developed an online list of Web sites related to the subject of this book. This site is updated regularly. Please use this link to access the list:
www.powerkidslinks.com/gosci/cyearth/